INTRODUCTION

When we work together as a team, we help each other and get things done. We are all good at different things, from remembering journeys to preparing meals and even building homes. It's the same with many animals. When they live with others, they thrive – together the group can teach the young, share food and warmth, and protect individuals from danger. Teamwork is powerful.

3

TEAM STARLING

Flocks of common starlings roost together, feed together, and fly together in groups that can number anything from a few dozen to a few thousand individuals. These social birds enjoy the benefits of even larger flock sizes in the winter months, when migrating birds arrive in huge numbers.

FACT FILE
Length: 20–23 cm (8–9 in)
Weight: 60–96 g (2–3½ oz)
Wingspan: 31–40 cm (12–16 in)

SUPER CHATTY

Common starlings are noisy birds. They love to chatter away and will use whirrs and whistles to talk to each other. Common starlings are also expert mimics – they can copy other bird calls.

Chatter

4

ANIMAL TEAMS

How amazing animals work
together in the wild

Illustrated by
CHARLOTTE MILNER

Written by **CAROLINE STAMPS**

DK

Penguin Random House

Senior Editor Carrie Love
Art Editor Polly Appleton
Managing Editor Penny Smith
Deputy Art Director Mabel Chan
Production Editor Dragana Puvacic
Production Controller Inderjit Bhullar
Jacket Designer Polly Appleton
Text by Caroline Stamps
Subject consultant Kim Dennis-Bryan

First published in Great Britain in 2022 by
Dorling Kindersley Limited
DK, One Embassy Gardens, 8 Viaduct Gardens,
London, SW11 7BW

The authorised representative in the EEA is
Dorling Kindersley Verlag GmbH. Arnulfstr. 124,
80636 Munich, Germany

Copyright © 2022 Dorling Kindersley Limited
A Penguin Random House Company
10 9 8 7 6 5 4 3 2 1
001–325874–Feb/2022

All rights reserved.
No part of this publication may be reproduced, stored in
or introduced into a retrieval system, or transmitted, in any
form, or by any means (electronic, mechanical, photocopying,
recording, or otherwise), without the prior written
permission of the copyright owner.

A CIP catalogue record for this book
is available from the British Library.
ISBN: 978-0-2415-2591-3

Printed and bound in China

For the curious
www.dk.com

MIX
Paper from
responsible sources
FSC™ C018179

This book was made with Forest
Stewardship Council™ certified
paper – one small step in DK's
commitment to a sustainable future.
For more information go to
www.dk.com/our-green-pledge

CONTENTS

FINDING FOOD TOGETHER

Common starlings often forage in large flocks, swooping down to look for insects. They also eat worms, fruits, and seeds. The young soon learn to copy their parents, poking their beak into the soil to find tasty snacks.

Worm

Male common starling

TRAVEL BUDDIES

Some common starlings migrate together from eastern Europe to western and southern Europe, when the weather gets cold.

SHIMMERING COLOUR

Juvenile common starlings have dull, brown feathers, but adults have summer plumage that is glossy green, blue, and purple with lots of white spots. Males have a blue base to their yellow beak, while females have a pink base on their beak.

5

LARGER GATHERINGS

In the autumn and winter months, common starlings form even larger flocks, swooping and diving together as if they are one creature. This is called a murmuration.

Murmuration

SAFETY AND WARMTH

A bird of prey will struggle to catch a single common starling that is flying with thousands of other birds in a murmuration. Common starlings roost together because their combined body heat as a group helps to keep them all warm.

6

BEDTIME TALKS

Before starlings settle down for the night, they chatter away to each other. The noise can be heard for quite a distance.

HAPPY HOME

Flocks often settle on treetops during the day, keeping a watchful eye for predators. They take to the skies at dusk, before finding less exposed areas, such as reedbeds, to roost in for the night.

FACT FILE
Length: 27–32 cm (10½–12½ in)
Tail length: 37–44 cm (14½–17 in)
Weight: 0.7–1 kg (1½–2 lb)

TEAM MONKEY

Black-capped squirrel monkeys live in humid forests in South America. They live together in large groups of up to 65 monkeys. A group of monkeys is called a troop. By living in a troop they are able to look out for each other.

SWINGING TAILS

Black-capped squirrel monkeys' tails are longer than their bodies. They cannot grip with their tails, but use them for balance.

Long tail

SLEEP AND PLAY

During the day the group searches for food in the forest. At night they often sleep huddled together among the trees, wrapping their long tails around their bodies for warmth.

LIFE AS A BABY

Baby black-capped squirrel monkeys are tiny – one would easily fit on your hand. Mothers have one infant, and for the first few weeks the infant clings to its mother before starting to explore further. Females in a troop help each other with childcare – a benefit of communal life.

Snack

TASTY SNACKS

As omnivores, black-capped squirrel monkeys eat plants and animals. Favourite foods are fruits and insects.

COMMUNICATION

A troop uses a number of
different sounds to keep
in touch with each other.
They purr, bark, squeal,
and squawk.

A WEE STORY

These monkeys wee on their hands and
spread this scent on their bodies.
Scientists think that males can tell from
the smell if a female wants to mate.

CHANGING SHAPE

Males become bigger around the shoulders
and arms when it's the mating season.

Ear

Hand

KEEPING CLEAN

Black-capped squirrel monkeys spend time grooming. They pick out hidden pests and seeds that might be caught in their coats, leaving their fur clean and fluffed up. They tend to self-groom using both their hands and feet, but mothers grooming their young use only their hands.

FACT FILE

Length: 87–130 cm (34–51 in)

Tail length: 35–52 cm (14–20 in)

Shoulder height: 66–81 cm (26–32 in)

Weight: Up to 62 kg (136½ lb)

TEAM WOLF

Grey wolves live in North America, Asia, and Europe. They group together in packs of up to 12 members. A pack is led by two parents who usually stay together for life. They are social animals and rely on the pack working together to survive.

A FAMILY GROUP

Packs of grey wolves are made up of a male, a female, and their pups. Packs may include youngsters up to three years old that haven't yet left to find their own territory.

12

BODY LANGUAGE

Grey wolves depend on body language to communicate with each other. Young and subordinate grey wolves will crouch down and roll over to show respect to dominant members of the pack.

A SAFE SPACE

In the early months, grey wolf cubs learn in an open area known as a "rendezvous site", where they play, eat, and sleep. They learn what to do by following their parents' lead, just like human children learn from the adults they live with.

Female wolf and pups

Male wolf

Wolf pup

A LARGE TERRITORY

Grey wolves roam huge areas in search of prey –
they may travel around 48 km (30 miles) a day on
a hunt for food, and more if necessary. They
communicate over large distances by howling.

CATCHING PREY

Once the pups reach the age of about four months,
they begin to follow their parents or older members
of the pack on hunts. Wolves tend to hunt large,
grazing animals, such as deer and bison.

COMMUNICATION IS KEY

Every grey wolf has a howl that is unique to them – in
the same way as human voices are all different.

BABY WOLVES

Grey wolves are born deaf and blind and rely on their mother's protection. The whole pack helps to raise and feed the pups. The pups learn through play fighting.

WHAT'S THAT SMELL?

Grey wolves have a sense of smell that is about 100 times better than a human's.

15

TEAM ANT

Leaf-cutter ants live and work in huge groups called colonies. Their underground nests often number millions of individuals. Within these colonies, each type of ant has a particular job to do.

I'LL DO THAT

Leaf-cutter ants have different roles. A queen, the largest ant, produces eggs to keep the colony thriving. The largest workers, known as soldiers, protect the colony. Other workers collect food. The smallest workers clean up, nurse the larvae, or tend vast areas known as gardens.

SUPER STRONG

The workers can carry things that are up to 50 times heavier than themselves. Loaded up, they return to the nest following a trail that can be many metres in length.

Worker ant

IT'S THAT WAY

If a worker finds a good source of food, it leaves an invisible scent trail that other ants can follow. Once established, some ants work to keep these trails clear of (small) debris. Team work makes for a faster supply chain!

POWERFUL JAWS

Leaf-cutter ant's jaws, known as mandibles, contain v-shaped, razor-sharp blades that cut through a leaf very easily. As leaf-cutter ants age, the blades become less sharp, so younger ants take over the leaf cutting while the older ants carry leaves back to the nest.

GROWING GARDENS

At the nest, the leaves are cut up and used to grow fungus "gardens", which provide food for the ants and for larvae. Small worker ants called minims look after these gardens with great care.

BEING EFFICIENT

Workers travel to and from their nest in single file. In doing so they move at the same speed and carry similar-sized loads.

SAFETY PROVIDERS

The colony is also protected by fierce soldier ants. They will bite attackers to defend the colony.

TUNNEL BUILDERS

Leaf-cutter ant nests are made up of hundreds of chambers that spread across 30 m (98 ft) or more. They are excavated mouthful by mouthful! The dirt is carried to the surface and used to build the mound.

Soldier ant

Tunnel entrance

Larva

BABYSITTING

Leaf-cutter ants develop from eggs into larvae, then to pupae before emerging as adults. In development they are also looked after by the small leaf-cutter workers known as minims.

TEAM SARDINE

A group of sardines is called a school. Pacific sardines are small, typically less than 20 cm (8 in) in length. They live in coastal waters of the Indo-Pacific and eastern Pacific oceans.

A MASS MOVEMENT

Between May and July, billions of sardines migrate along the east coast of South Africa. It is known as the great sardine run.

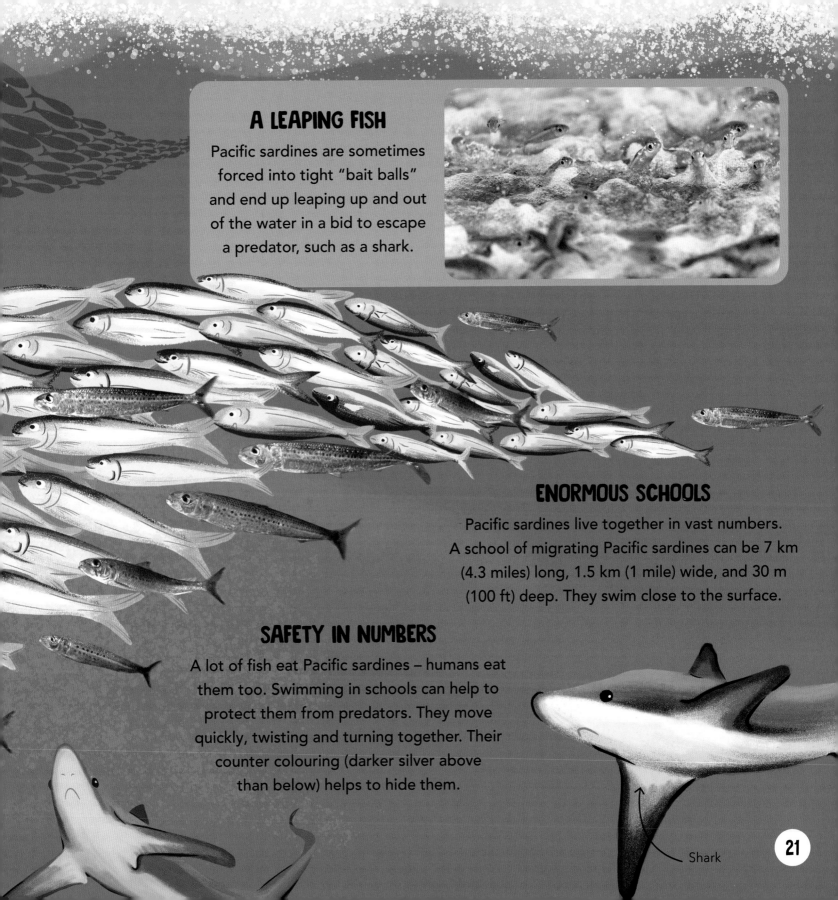

A LEAPING FISH

Pacific sardines are sometimes forced into tight "bait balls" and end up leaping up and out of the water in a bid to escape a predator, such as a shark.

ENORMOUS SCHOOLS

Pacific sardines live together in vast numbers. A school of migrating Pacific sardines can be 7 km (4.3 miles) long, 1.5 km (1 mile) wide, and 30 m (100 ft) deep. They swim close to the surface.

SAFETY IN NUMBERS

A lot of fish eat Pacific sardines – humans eat them too. Swimming in schools can help to protect them from predators. They move quickly, twisting and turning together. Their counter colouring (darker silver above than below) helps to hide them.

Shark

WHAT'S IN A NAME?

The name "sardine" is a general name, given to a number of small, silvery fish that are in the herring family. There are different types, including rainbow sardines and white sardines.

SEEKING FOOD

Pacific sardines largely feed on tiny organisms called plankton, many smaller in size than the head of a pin.

A CONTINUOUS STARE

Sardines have eyes with lenses and circular pupils, just like us, but they don't need eyelids. Their large eyes help sardines see better in low light conditions.

AN EXCESS OF EGGS

Pacific sardines lay about 100,000 eggs a year each. The majority will be eaten by fish and other creatures, including corals.

TEAM RABBIT

European rabbits live in Western Europe and North Africa in groups, or colonies. Their homes are called warrens. These underground networks of tunnels and chambers provide the rabbits with a safe place to escape from predators.

Buck

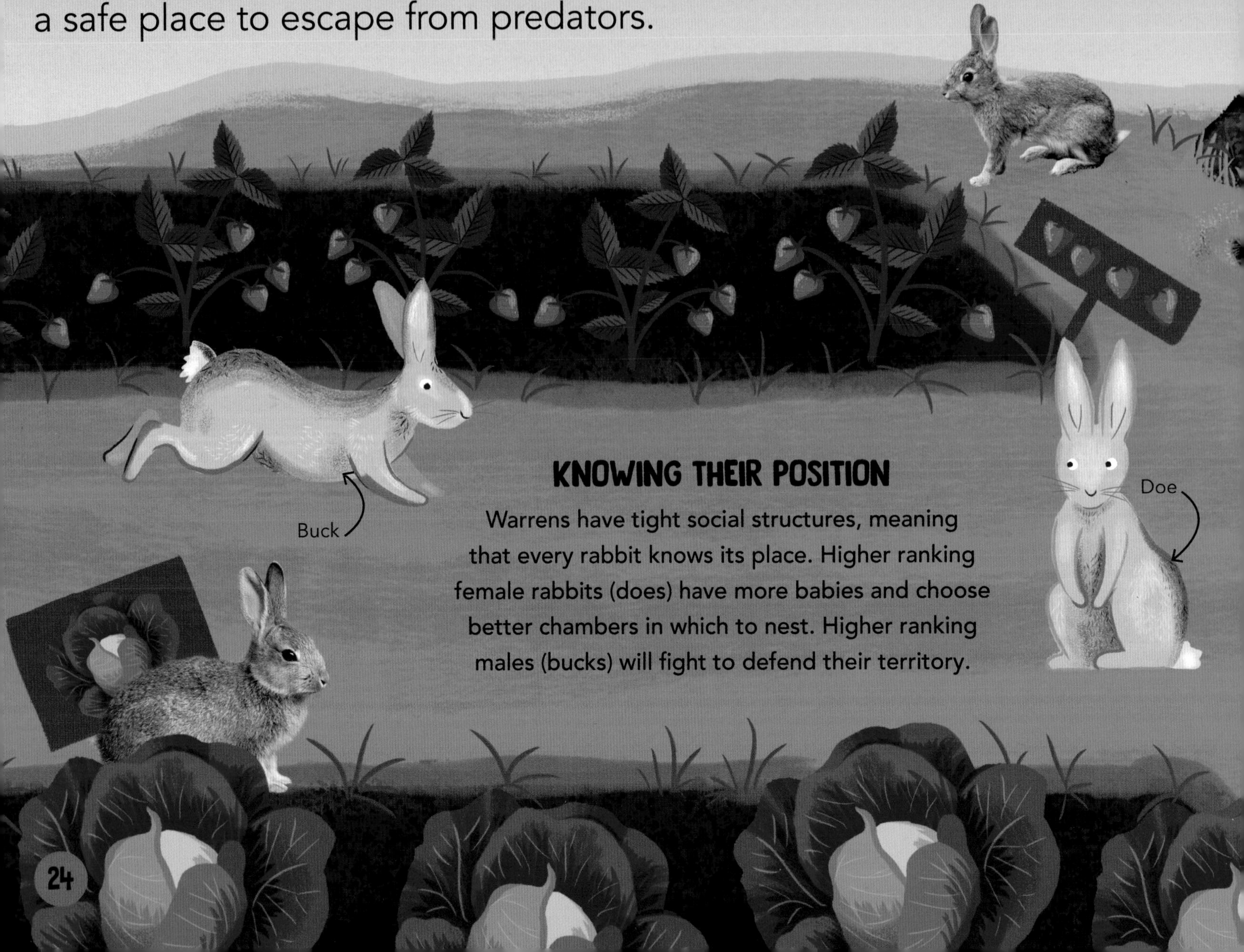

KNOWING THEIR POSITION

Warrens have tight social structures, meaning that every rabbit knows its place. Higher ranking female rabbits (does) have more babies and choose better chambers in which to nest. Higher ranking males (bucks) will fight to defend their territory.

Doe

BUNNY BUDDIES

Rabbits are constantly alert to danger. They depend on each other for reassurance and companionship. They should never be kept alone as pets.

Ears pricked to listen for danger

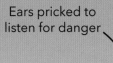

WATCH OUT!

Rabbits communicate using body language, touch, and their sense of smell. They sit up on their hind legs to see further away and if they sense danger will thump one or both hind feet.

POO POWER

Some plants that rabbits eat are hard to digest. These pass through their digestive system and emerge as soft droppings. Rabbits eat these as they are rich in nutrients.

DIGGING BURROWS

Female European rabbits do most of the digging. Once a rabbit finds a safe place it begins to dig with its front paws. It's a big job as warrens have more than one entrance.

Wild rabbits don't actually eat carrots. Pet rabbits can eat some as a treat.

TIDY WARREN

Up to 60 rabbits may live in a warren. They keep it very clean – rabbits don't leave droppings in their burrows.

Tunnel

26

Doe

Kittens cosy
in their nest

Chamber

A COSY CHAMBER

Baby rabbits, or kittens, sleep in chambers that the mother lines with grass and fur. There may be up to nine rabbits in a litter.

LOOKING AFTER THEIR YOUNG

A doe takes great care to look after its kittens as the babies are born blind, deaf, and almost hairless. However, they grow quickly. They suckle their mother's milk for a few minutes a day and snuggle up to each other for warmth. They open their eyes by day 10 and by day 18–20 they begin to leave the burrow. They are independent by day 30.

ANIMAL FACT FILE

COMMON STARLING

Location: Native to Europe, Southwest Asia, and northern Africa. Introduced to North America, South Africa, Australia, New Zealand, and Polynesia

Food: Insects, snails, worms, fruit, and seeds

Baby name: Chick **Number of babies:** Four to six eggs at a time

Habitat (home): Lowland, open areas with scattered trees

Group name: Flock of birds or murmuration of starlings

BLACK-CAPPED SQUIRREL MONKEY

Location: Peru, Bolivia, and upper Amazon of Brazil

Food: Fruit and small animals (insects, snails, small vertebrates, such as frogs)

Baby name: Infant or baby

Number of babies: One at a time

Habitat (home): Wet, humid forest

Group name: Troop or group

GREY WOLF

Location: North America, Europe, and Asia

Food: Deer, buffalo, rabbits, mice, and carrion

Baby name: Pup

Number of babies: About five in a litter

Habitat (home): Northern areas, ranging from tundra (treeless, vast cold, and flat areas) to deserts

Group name: Pack

LEAF-CUTTER ANT

Location: Central America and South America
Food: Fungus grown on rotten leaf pieces
Baby name: Larva
Number of babies: Thousands hatch every day
Habitat (home): Edges of tropical and scrub forests, and agricultural land
Group name: Colony or swarm

PACIFIC SARDINE

Location: Indo-Pacific and eastern Pacific
Food: Plankton
Baby name: Larva
Number of babies: Up to 45,000 eggs per spawning event
Habitat (home): Surface waters of the ocean
Group name: School

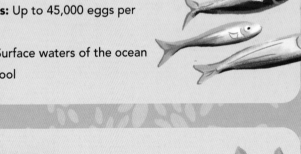

EUROPEAN RABBIT

Location: Western Europe and North Africa, introduced elsewhere
Food: Mainly grass and some flowering plants
Baby name: Kitten
Number of babies: About four in a litter
Habitat (home): Grassland and woodland
Group name: Colony

OTHER TEAMS

From large groups to small pairings, animals support each other in different ways. Female orangutans live with their young for up to 8 years. Dolphins help each other if a member of the pod gets injured, bees share information, and seahorse couples mate for life.

BOTTLENOSE DOLPHIN

Location: Oceans worldwide to around 45 degrees north and south of the equator
Food: Fish and squid
Baby name: Calf
Number of babies: One
Habitat (home): Temperate and tropical waters
Group name: Pod

WESTERN HONEYBEE

Location: Almost worldwide, except near the cold poles
Food: Nectar and pollen
Baby name: Larva
Number of babies: Up to 2,000 eggs laid by the queen each day
Habitat (home): Grassland, woodland, heath and moorland, agricultural land, and gardens
Group name: Swarm (out of hive), colony (in hive)

BORNEAN ORANGUTAN

Location: Island of Borneo in Southeast Asia

Food: Mainly fruit – also leaves and termites

Baby name: Infant or baby

Number of babies: One

Habitat (home): Tropical forests of Borneo

Group name: No group name, mother and infant live together for about eight years

YELLOW SEAHORSE

Location: Indian Ocean, Pacific Ocean

Food: Small crustaceans

Baby name: Fry

Number of babies: 20–1,000

Habitat (home): Shallow inshore waters of the Indo-Pacific

Group name: Herd

JAGUAR

Location: Arizona USA, Mexico, Central and South America as far as northern Argentina

Food: Wide ranging, including deer, capybaras, peccaries, birds, and monkeys

Baby name: Cub **Number of babies:** One to four

Habitat (home): Tropical forest and grassy swamp

Group name: No group name

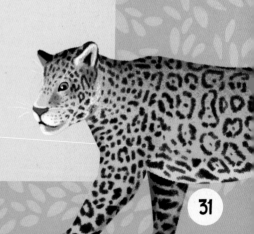

GLOSSARY

bait ball a school of fish forced to swim tightly together by predators

colony a large group of the same animal, such as ants, that live together

flock a group of birds that fly and search for food together

groom the act of an animal cleaning itself or another animal

larva an immature form of an insect or fish that hatches from an egg

mandible one half of an insects' jaws

migration seasonal movement of animals from one area to another

murmuration a huge number of starlings flying together in a co-ordinated way

omnivore a creature that eats plant- and animal-based food

plumage a bird's feathers

predator an animal that hunts, kills, and eats another animal

prey an animal that is hunted, killed, and eaten

pupa the stage in the life cycle of an insect when a larva transforms into an adult

subordinate in a wolf pack, the wolf that has a lower standing than the others

warren a network of tunnels and chambers dug out by rabbits

INDEX

ACKNOWLEDGEMENTS

Dorling Kindersley would like to thank the following people for their assistance in the preparation of this book: Jennifer Henshaw for proofreading, Sif Nørskov for additional photo research, Jagtar Singh for high resolution work on the images, and Rituraj Singh for picture research.

Picture Credits The publisher would like to thank the following for their kind permission to reproduce their photographs: (Key: a-above; b-below/bottom; c-centre; f-far; l-left; r-right; t-top) **1 Dreamstime.com:** Viniciussouza06 (t); Kevin Wells (cla). **5 Dreamstime.com:** Sander Meertins (cl, tl); Mikalay Varabey (tr). **6-7 Getty Images:** Bon Espoir Photography. **7 Alamy Stock Photo:** Bob Gibbons (bc). **Dreamstime.com:** Vasyl Helevachuk (cb). **8 Dreamstime.com:** Anna Podekova (bc); Suzanne Schoepe (cra). **Shutterstock.com:** Christopher Sharpe (tc). **8-9 Dreamstime.com:** Anna Podekova (Leaves). **9 Alamy Stock Photo:** Arco Images / Huetter, C. / Imagebroker (tr); Wild Sue (tc); Ger Bosma (clb). **Dreamstime.com:** Tristan Barrington (ca). **10 Dreamstime.com:** Mauro Rodrigues (cb). **10-11 Dreamstime.com:** Anna Podekova (Leaves). **11 Dreamstime.com:** Anna Kucherova (cl). **Getty Images / iStock:** Zoran Kolundzija (cr). **12 Alamy Stock Photo:** mauritius images GmbH (br). **12-13 Dreamstime. com:** Holly Kuchera (bc). **13 Alamy Stock Photo:** FLPA (tc); Janet Horton (bc). **Dreamstime.com:** Kjetil Kolbjornsrud (br). **14 Dreamstime.com:** Valeriy Kalyuzhnyy / Dragoneye (clb). **15 Dreamstime.com:** Anagram1 (b); Holly Kuchera (tc). **16-17 Dreamstime.com:** Viniciussouza06 (Ants); Kevin Wells (Leaves). **17 Dreamstime.com:** Sérgio Zacchi (cr). **18 Alamy Stock Photo:** Mark Moffett / Minden Pictures (tl). **Dreamstime.com:** Klaus Mohr (cb). **18-19 Dreamstime.com:** Viniciussouza06. **19 Alamy Stock Photo:** Christian Ziegler / Minden Pictures (tl). **Dreamstime.com:** Kevin Wells (cb). **20-21 Dreamstime.com:** Slowmotiongli. **21 naturepl.com:** Tony Wu (cra). **22-23 Dreamstime.com:** Slowmotiongli. **23 Dreamstime.com:** Aquanaut4 (tr). **24 Dreamstime.com:** Helen Davies (bl, cra). **25 Alamy Stock Photo:** Richard Dorn / imageBROKER (crb); Yukihiro Fukuda / Nature Picture Library (tl). **Dreamstime.com:** Helen Davies (cra); Mikelane45 (cl); Andrei Shupilo (c). **26 Dreamstime.com:** Isselee (crb); Wrangel (ca, ca/2). **27 Alamy Stock Photo:** Gary K Smith (tr). **Dreamstime.com:** Isselee (br). **28 123RF.com:** Mikalay Varabey (cla). **Dreamstime.com:** Simone Van Den Berg (cl). **Photolibrary:** DLILLC / Corbis (bl). **29 123RF.com:** Rudmer Zwerver (bl). **Dreamstime.com:** Slowmotiongli (cl); Kevin Wells (cla). **30 Dorling Kindersley:** Charlotte Milner. **Dreamstime.com:** Franky (cl); Oleksii Zelivianskyi (bl). **31 Dreamstime. com:** Alberto Carrera (cla); Elena Frolova (cl); Palko72 (bl)

All other images © Dorling Kindersley
For further information see: www.dkimages.com